In
1935 if you wanted to
read a good book, you needed
either a lot of money or a library card.
Cheap paperbacks were available, but their
poor production generally mirrored the quality
between the covers. One weekend that year,
Allen Lane, Managing Director of The Bodley Head,
having spent the weekend visiting Agatha Christie,
found himself on a platform at Exeter station trying to
find something to read for his journey back to London.
He was appalled by the quality of the material he had to
choose from. Everything that Allen Lane achieved from that
day until his death in 1970 was based on a passionate belief
in the existence of 'a vast reading public for *intelligent*
books at a low price'. The result of his momentous vision
was the birth not only of Penguin, but of the 'paperback
revolution'. Quality writing became available for the price of
a packet of cigarettes, literature became a mass medium
for the first time, a nation of book-borrowers became a
nation of book-buyers – and the very concept of book
publishing was changed for ever. Those founding
principles – of quality and value, with an overarching
belief in the fundamental importance of reading –
have guided everything the company has
done since 1935. Sir Allen Lane's
pioneering spirit is still very much alive
at Penguin in 2005. Here's to
the next 70 years!

MORE THAN A BUSINESS

'We decided it was time to end the almost customary half-hearted manner in which cheap editions were produced – as though the only people who could possibly want cheap editions must belong to a lower order of intelligence. We, however, believed in the existence in this country of a vast reading public for intelligent books at a low price, and staked everything on it'
Sir Allen Lane, 1902–1970

'The Penguin Books are splendid value for sixpence, so splendid that if other publishers had any sense they would combine against them and suppress them'
George Orwell

'More than a business ... a national cultural asset'
Guardian

'When you look at the whole Penguin achievement you know that it constitutes, in action, one of the more democratic successes of our recent social history'
Richard Hoggart

The View from Mount Improbable

RICHARD DAWKINS

PENGUIN BOOKS

PENGUIN BOOKS

Published by the Penguin Group
Penguin Books Ltd, 80 Strand, London WC2R 0RL, England
Penguin Group (USA) Inc., 375 Hudson Street, New York, New York 10014, USA
Penguin Group (Canada), 10 Alcorn Avenue, Toronto, Ontario, Canada M4V 3B2
(a division of Pearson Penguin Canada Inc.)
Penguin Ireland, 25 St Stephen's Green, Dublin 2, Ireland
(a division of Penguin Books Ltd)
Penguin Group (Australia), 250 Camberwell Road, Camberwell, Victoria 3124,
Australia (a division of Pearson Australia Group Pty Ltd)
Penguin Books India Pvt Ltd, 11 Community Centre,
Panchsheel Park, New Delhi – 110 017, India
Penguin Group (NZ), cnr Airborne and Rosedale Roads, Albany,
Auckland 1310, New Zealand (a division of Pearson New Zealand Ltd)
Penguin Books (South Africa) (Pty) Ltd, 24 Sturdee Avenue,
Rosebank 2196, South Africa

Penguin Books Ltd, Registered Offices: 80 Strand, London WC2R 0RL, England

www.penguin.com

Climbing Mount Improbable first published by Viking 1996
This extract published as a Pocket Penguin 2005

1

Grateful acknowledgement is given to A. P. Watt Ltd on behalf of Anne and
Michael Yeats for permission to reprint an extract from 'The Cat and
the Moon', from *The Collected Poems of W. B. Yeats*

Set in 10.5/12.5pt Monotype Dante
Typeset by Palimpsest Book Production Limited
Polmont, Stirlingshire
Printed in England by Clays Ltd, St Ives plc

I have just listened to a lecture in which the topic for discussion was the fig. Not a botanical lecture, a literary one. We got the fig in literature, the fig as metaphor, changing perceptions of the fig, the fig as emblem of pudenda and the fig leaf as modest concealer of them, 'fig' as an insult, the social construction of the fig, D. H. Lawrence on how to eat a fig in society, 'reading fig' and, I rather think, 'the fig as text'. The speaker's final *pensée* was the following. He recalled to us the Genesis story of Eve's tempting Adam to eat of the fruit of the tree of knowledge. Genesis doesn't specify, he reminded us, which fruit it was. Traditionally, people take it to be an apple. The lecturer suspected that actually it was a fig, and with this piquant little shaft he ended his talk.

This kind of thing is the stock-in-trade of a certain kind of literary mind, but it provokes me to irritated literal-mindedness. The speaker obviously knew that there never was a Garden of Eden, never a tree of knowledge of good and evil. So what was he actually trying to say? I suppose he had a vague feeling that 'somehow', 'if you will', 'at some level', 'in some sense', 'if I may put it this way' it is somehow 'right' that the fruit in the story 'should' have been a fig. But enough of this foolery. It is not that we should be literalist and Gradgrindian, but our elegant lecturer was *missing* so much. There is genuine paradox and real poetry lurking in the fig, with subtleties to exercise an inquiring mind and wonders to uplift an aesthetic one. In *Climbing Mount*

Improbable, the book from which this extract is taken, I try to move to a position where I can tell the true story of the fig. But the fig story is only one out of millions that all have the same Darwinian grammar and logic – albeit the fig story is among the most satisfyingly intricate in all evolution. To anticipate the central metaphor of the book, the fig tree stands atop one of the highest peaks on the *massif* of Mount Improbable. But peaks as high the fig's are best conquered at the end of the expedition. Before that there is much that needs to be said, a whole vision of life that needs to be developed and explained, puzzles that need to be solved and paradoxes that must be disarmed.

The fig's story is, at the deepest level, the same story as for every other living creature on this planet. Though they differ in surface detail, all are variations on the theme of DNA and the thirty million ways by which it propagates itself. In *Climbing Mount Improbable* we look at spider webs – at the bewildering, though unconscious, ingenuity with which they are made and how they work. We reconstruct the slow, gradual evolution of wings and of elephant trunks. We see that 'the' eye, legendarily difficult though its evolution sometimes seems, has actually evolved at least forty and perhaps sixty times independently all around the animal kingdom. We program computers to assist our imagination in moving easily through a gigantic museum of all the countless creatures that have ever lived and died, and their even more numerous imaginary cousins who have never been born. We wander the paths of Mount Improbable, admiring its vertical precipices from afar, but always restlessly seeking the gently graded slopes on the other side. And finally we return to the fig, fit and trained by our experience on the mountain to tackle that most daunting of Darwinian puzzles.

*

Mount Improbable rears up from the plain, lofting its peaks dizzily to the rarefied sky. The towering, vertical cliffs of Mount Improbable can never, it seems, be climbed. Dwarfed like insects, thwarted mountaineers crawl and scrabble along the foot, gazing hopelessly at the sheer, unattainable heights. They shake their tiny, baffled heads and declare the brooding summit for ever unscalable.

Our mountaineers are too ambitious. So intent are they on the perpendicular drama of the cliffs, they do not think to look round the other side of the mountain. There they would find not vertical cliffs and echoing canyons but gently inclined grassy meadows, graded steadily and easily towards the distant uplands. Occasionally the gradual ascent is punctuated by a small, rocky crag, but you can usually find a detour that is not too steep for a fit hill-walker in stout shoes. If you are prepared to take your time, you can reach the summit of Mount Improbable in a leisurely stroll. The sheer height of the peak doesn't matter, so long as you don't try to scale it in a single bound. Locate the mildly sloping path and, if you have unlimited time, the ascent is only as formidable as the next step.

One of Britain's most famous physical scientists, Sir Fred Hoyle (incidentally the author of *The Black Cloud*, which must be among the best science-fiction novels ever written) raised the question of how complicated molecules such as enzymes might evolve. Enzymes work in cells rather like exceedingly numerous machine tools for molecular mass-production. Their efficacy depends upon their three-dimensional shape, their shape depends upon their coiling behaviour, and their coiling behaviour depends upon the sequence of amino-acids which link up in a chain to make them. This exact sequence really matters. Could it come about by chance?

Hoyle said no, and he was right. There is a fixed repertoire of amino acids available, twenty. A typical enzyme is a chain of several hundred links drawn from the twenty. An elementary calculation shows that the probability that any particular sequence of, say, 100 amino acids will spontaneously form is 1 in 20 x 20 x 20 . . . 100 times, or 1 in 20^{100}. This is an inconceivably large number, far greater than the number of fundamental particles in the entire universe. His co-author and fellow astrophysicist, Chandra Wickramasinghe, quoted Sir Fred as saying that the spontaneous formation by 'chance' of a working enzyme is like a hurricane blowing through a junkyard and spontaneously having the luck to put together a Boeing 747. What Hoyle and Wickramasinghe overlooked is that Darwinism is *not* a theory of random chance. Natural selection, to be sure, needs (random) mutation as the original source of the variation upon which it works, but natural selection itself is the very antithesis of a chance process.

It is grindingly, creakingly, crashingly obvious that, if Darwinism were really a theory of chance, it couldn't work. You don't need to be a mathematician or physicist to calculate that an eye or a haemoglobin molecule would take from here to infinity to self-assemble by sheer higgledy-piggledy luck. Far from being a difficulty peculiar to Darwinism, the astronomic improbability of eyes and knees, enzymes and elbow joints and all the other living wonders is precisely the problem that *any* theory of life must solve, and that Darwinism uniquely *does* solve. It solves it by breaking the improbability up into small, manageable parts, smearing out the luck needed, going round the back of Mount Improbable and crawling up the gentle slopes, inch by million-year inch.

The height of Mount Improbable stands for the combination of perfection and improbability that is epitomized in eyes and enzyme molecules (and intelligences capable of

4

designing them – Fred Hoyle should see God as the ultimate Boeing 747). To say that an object like an eye or a protein molecule is improbable means something rather precise. The object is made of a large number of parts arranged in a very special way. The number of possible ways in which those parts could have been arranged is exceedingly large. In the case of a protein molecule we can actually calculate that large number. In the case of the eye we can't do the equivalent calculation without fabricating lots of assumptions, but we can intuitively see that it is going to come to another stupefyingly large number. The actual, observed arrangement of parts is improbable in the sense that it is only one arrangement among trillions of possible arrangements.

The Darwinian explanation for why living things are so good at doing what they do is very simple. They are good because of the accumulated wisdom of their ancestors. But it is not wisdom that they have learned or acquired. It is wisdom that they chanced upon by lucky random mutations, wisdom that was then selectively, nonrandomly, recorded in the genetic database of the species. In each generation the amount of luck was not very great: small enough to be believable. But, because the luck has been accumulated over so many generations (the gradual slope at the back of the mountain), we are very impressed by the apparent improbability of the end product (the sheer cliff at the front).

The message from the mountain is threefold. First, there can be no sudden leaps upward – no precipitous increases in ordered complexity. Second, there can be no going downhill – species can't get worse as a prelude to getting better. Third, there may be more than one peak – more than one way of solving the same problem, all flourishing in the world.

Take any part of any animal or plant, and it is a sensible

question to ask how that part has been formed by gradual transformation from some part of an earlier ancestor. We are ready to take up the creationists' favourite target, and the star stumbling block for would-be believers in evolution, perched precariously on the summit of the most formidable cliff Mount Improbable seems to offer: the eye.

All animals have to deal with their world, and the objects in it. They walk on objects, crawl under them, avoid crashing into them, pick them up, eat them, mate with them, run away from them. Back in the geological dawn when evolution was young, animals had to make physical contact with objects before they could tell that those objects were there. What a bonanza of benefit was waiting for the first animal to develop a remote-sensing technology: awareness of an obstacle before hitting it; of a predator before being seized; of food that wasn't already within reach but could be anywhere in the large vicinity. What might this high technology be?

The sun provided not only the energy to drive the chemical cogwheels of life. It also offered the chance of a remote guidance technology. It pummelled every square millimetre of Earth's surface with a fusillade of photons: tiny particles travelling in straight lines at the greatest speed the universe allows, criss-crossing and ricocheting through holes and cracks so that no nook escaped, every cranny was sought out. Because photons travel in straight lines and so fast, because they are absorbed by some materials more than others and reflected by some materials more than others, and because they have always been so numerous and so all-pervading, photons provided the opportunity for remote-sensing technologies of enormous accuracy and power. It was necessary only to detect photons and – more difficult – distinguish the directions from which they came. Would the opportunity be taken up? Three billion years later you know the answer, for you can see these words.

Darwin famously used the eye to introduce his discussion on 'Organs of extreme perfection and complication':

To suppose that the eye, with all its inimitable contrivances for adjusting the focus to different distances, for admitting different amounts of light, and for the correction of spherical and chromatic aberration, could have been formed by natural selection, seems, I freely confess, absurd in the highest possible degree.

It is possible that Darwin was influenced by his wife Emma's difficulties with this very point. Fifteen years before the *Origin* he had written a long essay outlining his theory of evolution by natural selection. He wanted Emma to publish it in the event of his death and he let her read it. Her marginalia survive, and it is particularly interesting that she picked out his suggestion that the human eye 'may *possibly* have been acquired by gradual selection of slight but in each case useful deviations'. Emma's note here reads, 'A great assumption/ E.D.' Long after the *Origin* was published Darwin confessed, in a letter to an American colleague: 'The eye, to this day, gives me a cold shudder, but when I think of the fine known gradations, my reason tells me I ought to conquer the cold shudder.' Darwin's doubts are much quoted by creationists – without the subsequent passages which show that, for him, the posing of a difficult problem was a rhetorical device to set off his triumphant solution of it. Darwin always saw a difficulty as a challenge to go on thinking, not a welcome excuse to give up.

When we speak of 'the' eye, by the way, we are not doing

justice to the problem. It has been authoritatively estimated that eyes have evolved no fewer than forty times and perhaps more than sixty times, independently in various parts of the animal kingdom. In some cases these eyes use radically different principles. Nine distinct principles have been recognized among the forty to sixty independently evolved eyes. I'll mention some of the nine eye types – which we can think of as nine distinct peaks in different parts of Mount Improbable's *massif* – as I go on.

Accepting the limitations of the metaphor of Mount Improbable, let's go right down to the bottom of the vision slopes. Here we find eyes so simple that they scarcely deserve to be recognized as eyes at all. It is better to say that the general body surface is slightly sensitive to light. This is true of some single-celled organisms, some jellyfish, starfish, leeches and various other kinds of worms. Such animals are incapable of forming an image, or even of telling the direction from which light comes. All that they can sense (dimly) is the presence of (bright) light, somewhere in the vicinity. Weirdly, there is good evidence of cells that respond to light in the genitals of both male and female butterflies. These are not image-forming eyes but they can tell the difference between light and dark and they may represent the kind of starting point that we are talking about when we speak of the remote evolutionary origins of eyes. Nobody seems to knows how the butterflies use them, not even William Eberhard, whose diverting book, *Sexual Selection and Animal Genitalia*, is my source for this information.

If we think of the plain below Mount Improbable as peopled by ancestral animals that were totally unaffected by light, the non-directional light-sensitive skins of starfish and leeches (and butterfly genitals) are just a little way up the lower slopes, where the mountain path begins. It is not difficult to find the path. Indeed it may be that the 'plain' of total insensitivity to light has always been small. It might be that living cells are more or less bound to be somewhat affected by light – a possibility that makes the butterfly's light-sensitive genitals seem less strange. A light ray consists of a straight stream of photons. When a photon hits a molecule of some coloured substance it may be stopped in its tracks and the molecule changed into a different form of the same molecule. When this happens some energy is released. In green plants and green bacteria, this energy is used to build food molecules, in the set of processes called photosynthesis. In animals the energy may trigger a reaction in a nerve, and this constitutes the first step in the process called seeing, even in animals lacking organs that we would recognize as eyes. Any of a wide variety of coloured pigments will do, in a rudimentary way. Such pigments abound, for all sorts of purposes other than trapping light. The first faltering steps up the slopes of Mount Improbable would have consisted in the gradual improvement of pigment molecules. There is a shallow, continuous ramp of improvement – easy to climb in small steps.

This lowland ramp pushed on up towards the evolution of the living equivalent of the photocell, a cell specialized for capturing photons with a pigment, and translating their impact into nerve impulses. I shall continue to use the word photocell for those cells in the retina (in ourselves they are called rods and cones) which are specialized for capturing photons. The trick that they all use is to increase the number of layers of pigment available to capture photons. This is important because a photon is very likely to pass straight through any one layer of pigment and come out the other side, unscathed. The more layers of pigment you have, the greater the chance of catching any one photon. Why should it matter how many photons are trapped and how many get through? Aren't there always plenty of photons to spare? No, and the point is fundamental to our understanding of the design of eyes. There is a kind of economics of photons, an economics as mean-spirited as human monetarist economics and involving inescapable trade-offs.

Before we even get into the interesting economic trade-offs, there can be no doubt that in absolute terms photons are in short supply at some times. On a crisp, starry night in 1986 I woke my two-year-old daughter Juliet and carried her, wrapped in blankets, out into the garden where I pointed her sleepy face towards the published location of Halley's Comet. She didn't take in what I was saying, but I stubbornly whispered into her ear the story of the comet and the certainty that I could never see it again but that she might when she was seventy-eight. I explained that I had woken her so that she'd be able to tell her grandchildren in 2062 that she had seen the comet before, and perhaps she'd remember her father for his quixotic whim in carrying her out into the night to show it to her (I may even have whispered the words 'quixotic' and 'whim' because small children like the sound of words they don't know, carefully articulated).

Probably some photons from Halley's Comet did indeed touch Juliet's retinas that night in 1986 but, to be truthful, I had a hard time convincing myself that I could see the comet. Sometimes I seemed to conjure a faint, greyish smear at approximately the right place. At other times it melted away. The problem was that the number of photons falling on our retinas was close to zero.

Photons arrive at random times, like raindrops. When it

is really raining properly we are in no doubt of the fact and wish our umbrella hadn't been stolen. But when rain starts gradually, how do we decide the exact moment when it begins? We feel a single drop and look up apprehensively, unconvinced until a second or a third drop arrives. When rain is spitting infrequently like this, one person may say that it is raining while his companion denies it. The drops can fall infrequently enough to hit one person a minute before his companion registers a hit. To be really convinced that there is light, we need the photons to patter on our retinas at an appreciably high rate. When Juliet and I gazed in the general direction of Halley's Comet, photons from the comet were probably hitting individual photocells on our retinas at the fantastically slow rate of about one every 40 minutes. This means that any one photocell could be saying 'Yes there is light there' while the vast majority of its neighbouring photocells did not. The only reason I received any sensation at all of a comet-shaped object was that my brain was summing up the verdicts of hundreds of photocells. Two photocells capture more photons than one. Three capture more than two, and so on up the slope of Mount Improbable. Advanced eyes like ours have millions of photocells densely packed like pile in a carpet, and each one of them is set up to capture as many photons as possible.

Photocells on their own just tell an animal whether there is light or not. The animal can tell the difference between night and day, and can tell when a shadow falls which might, for example, portend a predator. The next step of improvement must have been the acquisition of some rudimentary sensitivity to direction of light and direction of movement of, say, a menacing shadow. The minimal way of achieving this is to back the photocells with a dark screen on one side only. A transparent photocell without a dark screen receives light from all directions and cannot tell where light is coming from. An animal with only one photocell in its head can steer towards, or away from, light, provided the photocell is backed by a screen. A simple recipe for doing this is to swing the head like a pendulum from side to side; if the light intensity on the two sides is unbalanced, change direction until it is balanced. There are some maggots that follow this recipe for steering directly away from light.

But swinging your head from side to side is a rudimentary way of detecting the direction of light, fit for the lowest slopes of Mount Improbable. A better way is to have more than one photocell pointing in different directions, each one backed by a dark screen. Then by comparing the rates of photon rain on the two cells you can make inferences about direction of light. If you have a whole carpet of photocells, a better way still is to bend the carpet, with its backing screen, into a curve, so that the photocells on different parts of the curve are pointing in systematically different directions. A convex curve can give rise, eventually, to the sort of 'compound eye' that insects have, and I'll return to this. A concave curve is a cup and it gives rise to the other main kind of eye, the camera eye like our own. Photocells in different parts of a cup will fire when light is coming from different directions, and the more cells there are the finer-grained will be the discrimination.

By keeping track of which photocells are firing and which are not, the brain can detect the direction from which the light is coming. From the point of view of climbing Mount Improbable, what matters is that there is a continuous evolutionary gradation – a smooth incline up the mountain – connecting animals with a flat carpet of photocells to animals with a cup. Cups can get gradually deeper or gradually shallower, by continuous slow degrees. The deeper the cup, the greater the ability of the eye to discriminate light coming from different directions. On the mountain, no steep precipices have to be leapt.

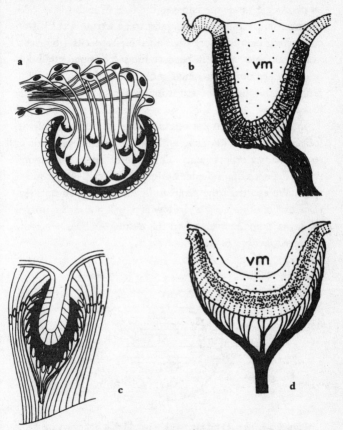

Figure 1 Cup eyes from around the animal kingdom: (a) flat-worm; (b) bivalve mollusc; (c) polychaet worm; (d) limpet.

Cup eyes like this are common in the animal kingdom. The eyes in Figure 1 have probably evolved their cup shape independently. This is particularly clear in the case of the flatworm eye which betrays its separate origin by keeping its photocells *inside* the cup. On the face of it this is an odd arrangement – the light rays have to penetrate a thicket of connecting nerves before they hit the photocells – but let's not be snobbish about it, because the same apparently poor design mars our own much more sophisticated eyes. I'll return to this and show that it isn't really such a bad idea as it seems.

In any case, a cup eye on its own is far from capable of forming what we humans, with our excellent eyes, would recognize as a proper image. Our kind of image-formation, which depends upon the lens principle, needs a little explanation. We approach the problem by asking why an unaided carpet of photocells, or a shallow cup, will not see an image of, say, a dolphin, even when the dolphin is conspicuously displayed in front of it.

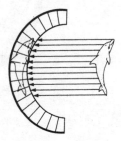

Figure 2 How eyes do not work – would that light rays were so obliging!

If light rays behaved as in Figure 2, everything would be easy, and an image of the dolphin, the right way up, would appear on the retina. Unfortunately they don't. To be more precise, there are rays that do exactly what I have drawn in the picture. The trouble is that these are swamped by any number of rays going in every other direction at the same time. Every bit of the dolphin sends a ray to every point on the retina. And not just every bit of the dolphin, but every bit of the background and of everything else in the scene. You can think of the result as an infinite number of dolphin images, in every possible position on the surface of the cup and every possible way up and way round. But what this amounts to, of course, is no image at all, just a smooth spreading of light over the whole surface (Figure 3).

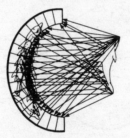

Figure 3 Light rays from everywhere go everywhere and no image is seen. An infinite number of dolphin images clash with each other, and nothing is clearly seen.

We have diagnosed the problem. The eye is seeing too much: an infinity of dolphins instead of only one. The obvious solution is to subtract: cut out all these dolphin images except one. It wouldn't matter which one, but how to get rid of all the rest? One way is to trudge on up the same slope of Mount

Improbable as gave us the cup, steadily deepening and enclosing the cup until the aperture has narrowed to a pinhole. Now the vast majority of rays are prevented from entering the cup at all. The minority that remain are just those rays that form a small number of similar images – upside down – of the dolphin (Figure 4). If the pinhole becomes extremely small, all the blurring disappears and a single, sharp picture of the dolphin remains (actually, extremely small pinholes introduce a new kind of blurring, but we'll forget about that for a moment). You can think of the pinhole as an image filter, removing all but one of the bewildering visual cacophony of dolphins.

The pinhole effect is just an extreme version of the cup effect that we have already met as an aid to telling the

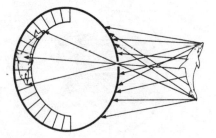

Figure 4 Principle of the pinhole eye. Most of the chaotic dolphin images are cut out. Only one (inverted) gets through the pinhole.

direction of light. It belongs only a bit farther up the same slope of Mount Improbable, and there are no sharp precipices between. There is no difficulty in a pinhole eye's evolving from a cup eye, and no difficulty in a cup eye's evolving from a flat sheet of photocells. The slope up the mountain from flat carpet to pinhole is gradual and easily climbable all the way. Climbing it represents a progressive knocking out of conflicting images until, at the peak, only one is left.

Pinhole eyes in varying degrees are, indeed, scattered around the animal kingdom. The most thoroughgoing pinhole eye is that of the enigmatic mollusc *Nautilus* (Figure 5a), related to the extinct ammonites (and a more distant relative of an octopus but with a coiled shell). Others, such as the eye of a marine snail in Figure 5b, are perhaps better described as deep cups than true pinholes. They illustrate the smoothness of this particular gradient up Mount Improbable.

A first thought suggests that the pinhole eye ought to work rather well, provided you make the pinhole small enough. If you make the pinhole almost infinitely small, you might think that you'd get an almost infinitely perfect image by cutting out the vast majority of competing, interfering images. But now two new snags arise. One is diffraction. I deferred talking about it just now. It is a blurring problem that results from the fact that light behaves like waves, which can interfere with each other. This blurring gets worse when the pinhole is very small. The other snag with a small pinhole recalls the hard trade-offs of our 'photon economy'. When the pinhole is small enough to make a sharp image, it necessarily follows that so little light gets through the hole that you can see the object well only if it is illuminated by an almost unattainably bright light. At normal lighting-levels not enough photons get through the pinhole for the eye to be certain what it is seeing. With a tiny pinhole we have a version of the Halley's Comet problem. You can combat this by opening out the pinhole again. But now you are back

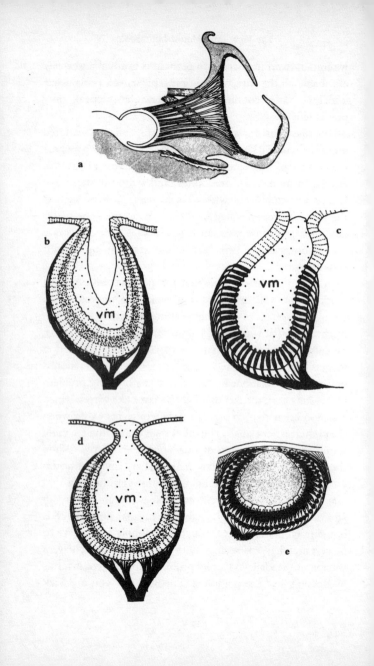

where you were with a confusion of competing 'dolphins'. The photon economy has brought us to an impasse on this particular foothill of Mount Improbable. With the pinhole design you can have a sharpish but dark image, or a bright-ish but fuzzy one. You cannot have both. Such trade-offs are the stock-in-trade of economists, which is why I coined the notion of an economy of photons. But is there no way to achieve a bright and yet simultaneously sharp image? Fortunately there is.

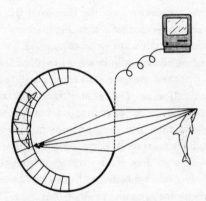

Figure 6 A complicated and expensive approach to the problem of forming an image that is both sharp and bright: the 'computed lens'.

Figure 5 (opposite) A range of invertebrate eyes that illustrate approaches to the formation of crude but effective images:
(a) *Nautilus*'s pinhole eye; (b) marine snail; (c) bivalve mollusc; (d) abalone; (e) ragworm.

First, think of the problem computationally. Imagine that we broaden the pinhole out to let in a nice lot of light. But instead of leaving it as a gaping hole, we insert a 'magic window', a masterpiece of electronic wizardry embedded in glass and connected to a computer. The property of this computer-controlled window is the following. Light rays, instead of passing straight through the glass, are bent through a cunning angle. This angle is carefully calculated by the computer so that all rays originating from a point (say the dolphin's nose) are bent to converge on a corresponding point on the retina. I've drawn only the rays from the dolphin's nose, but the magic screen, of course, has no reason to favour any one point and does the same calculation for all other points as well. So all rays originating at the tail of the dolphin are bent so that they converge on a corresponding tail point on the retina, and so on. The result of the magic window is that a perfect image of the dolphin appears on the retina. But it is not dark like the image from a tiny pinhole, because lots of rays (which means torrents of photons) converge from the nose of the dolphin, lots of rays converge from the tail of the dolphin, lots of rays converge from every point on the dolphin to their own particular point on the retina. The magic window has all the advantages of a pinhole, without its great disadvantage.

It's all very well to conjure up a so-called 'magic window' out of imaginative thin air. But isn't it easier said than done? Think what a complicated calculation the computer attached to the magic window is doing. It is accepting millions of light rays, coming from millions of points out in the world. Every point on the dolphin is sending millions of rays at millions of angles to different points on the surface of the magic window. The rays are criss-crossing one another in a bewildering spaghetti junction of straight lines. The magic window with its associated computer has to deal with each of these millions of rays in turn and calculate its own particular angle, through which it must be precisely turned. Where is this wonderful computer to come from, if not from a complicated miracle? Is this where we meet our Waterloo: an inevitable precipice in our ascent up Mount Improbable?

Remarkably, the answer is no. The computer in the diagram is just an imaginary creation to emphasize the *apparent* complexity of the task if you look at it in one way. But if you approach the problem in another way the solution turns out to be ludicrously easy. There is a device of preposterous simplicity which happens to have exactly the properties of our magic window, but with no computer, no electronic wizardry, no complication at all. That device is the lens. You don't need a computer because the calculations need never be done explicitly at all. The apparently complicated calculations of millions of ray angles are all taken care of, automatically and without fuss, by a curved blob of transparent material. I'll take a little time to explain how lenses work, as a prelude to showing that the evolution of the lens wouldn't have been very difficult.

It is a fact of physics that light rays are bent when they pass from one transparent material into another transparent material. The angle of bending depends upon which two materials they happen to be, because some substances have a greater refractive index – a measure of the capacity to bend light – than others. If we are talking about glass and water, the angle of bending is slight because the refractive index of water is nearly the same as that of glass. If the junction is between glass and air, the light is bent through a bigger angle because air has a relatively low refractive index. At the junction between water and air, the angle of bending is substantial enough to make an oar look bent.

Angles of refraction are easily calculated when the boundaries between surfaces are flat and parallel. But of course there is no reason why a blob of transparent material should have neatly parallel sides. Depending upon the angle at the surface of the blob, a ray can be sent off in any direction you choose. If the blob is covered with facets at lots of different angles, a set of rays can be sent off in lots of different directions. And if the blob is curved convexly on one or both of its sides, it will be a lens: the working equivalent of our magic window.

Transparent materials are not particularly rare in nature. Air and water, two of the commonest substances on our planet, are both transparent. So are many other liquids. So are some crystals if their surface is polished, for instance by wave action in the sea, to remove surface roughness. Imagine a pebble of some crystalline material, worn into a random shape by the waves. Light rays from a single source are bent in all sorts of directions by the pebble, depending upon the angles of the pebble's surfaces. Pebbles come in all sorts of shapes. Quite commonly they are convex on both sides. What will this do to light rays from a particular source like a light bulb?

When the rays emerge from a pebble with vaguely convex sides they will tend to converge. Not to a neat, single point such as would reconstruct a perfect image of the light bulb like our hypothetical 'magic window'. That would be too much to hope. But there is a definite tendency in the right direction. Any quartz pebble whose weathering happened to make it smoothly curvaceous on both sides would serve as a 'magic window', a true lens capable of forming images which, though far from sharp, are much brighter than a pinhole could produce. Pebbles worn by water usually are, as a matter of fact, convex on both sides. If they happened to be made of transparent material many of them would constitute quite serviceable, though crude, lenses.

A pebble is just one example of an accidental, undesigned object which can happen to work as a crude lens. There are others. A drop of water hanging from a leaf has curved edges. It can't help it. Automatically, without further design from us, it will function as a rudimentary lens. Liquids and gels fall automatically into curved shapes unless there is some force, such as gravity, positively opposing this. This will often mean that they cannot help functioning as lenses. The same is often true of biological materials. A young jellyfish is both lens-shaped and beautifully transparent. It works as a tolerably good lens, even though its lens properties are never actually used in life and there is no suggestion that natural selection has favoured its lens-like properties. The transparency probably is an advantage because it makes it hard for enemies to see, and the curved shape is an advantage for some structural reason having nothing to do with lenses.

Here are some images I projected onto a screen using various crude and undesigned image-forming devices. Figure 7a shows a large letter A, as projected on a sheet of paper at the back of a pinhole camera (a closed cardboard box with a hole in one side). You probably could scarcely read it if you weren't told what to expect, even though I used a very bright light to make the image. In order to get enough light to read it at all, I had to make the 'pin' hole quite large, about a centimetre across. I might have sharpened the image by narrowing the pinhole, but then the film would not have registered it – the familiar trade-off we have already discussed.

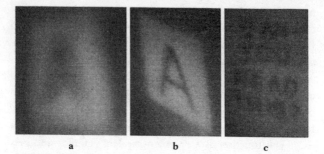

a   b   c

Figure 7 Images seen through various makeshift holes and crude, makeshift lenses: (a) a plain pinhole; (b) a sagging polythene bag filled with water; (c) a round wine goblet filled with water.

Now see what a difference even a crude and undesigned 'lens' makes. For Figure 7b the same letter A was again projected through the same hole onto the back wall of the same cardboard box. But this time I hung a polythene bag filled with water in front of the hole. The bag was not designed to be particularly lens-shaped. It just naturally hangs in a curvaceous shape when you fill it with water. Figure 7c ('CAN YOU READ THIS?') was made with the same cardboard box and hole, but this time a round wine goblet filled with water was placed in front of the hole instead of a sagging bag. Admittedly the wine glass is a man-made object, but its designers never intended it to be a lens and they gave it its shape for other reasons. Once again, an object that was not designed for the purpose turns out to be an adequate lens.

Of course, polythene bags and wine glasses were not available to ancestral animals. I am not suggesting that the evolution of the eye went through a polythene bag stage, any more than it went through a cardboard box stage. The point about the polythene bag is that, like a raindrop or a jellyfish or a rounded quartz crystal, it was not designed as a lens. It spontaneously takes on a lens-like shape for some other reason which happens to be influential in nature.

It is not difficult, then, for rudimentary lens-like objects to come into existence spontaneously. Any old lump of half-way transparent jelly need only assume a curved shape (there are all sorts of reasons why it might) and it will immediately confer at least a slight improvement on a simple cup or pinhole. Slight improvement is all that is required to inch up the lower slopes of Mount Improbable. What might the intermediates have looked like? Look back at Figure 5, and once again I must stress that these animals are all modern and must not be thought of as an actual ancestral series. Notice that the cup in Figure 5b (marine snail) has a lining of transparent jelly, the 'vitreous mass' (vm) which perhaps serves to protect the sensitive photocells from the raw sea water which flows freely through the aperture into the cup. That purely protective vitreous mass has one of the necessary qualities of a lens – transparency – but it lacks the correct curvature and it needs thickening up. Now look at Figures 5c, 5d and 5e, eyes from a couple of clam species, an abalone and a worm. In addition to providing yet more examples of cups and intermediates between cups and pinholes, all these eyes show greatly thickened vitreous masses. Vitreous masses, of varying degrees of shapelessness, are ubiquitous in the animal kingdom. As a lens, none of those splodges of jelly would move Mr Zeiss or Mr Nikon to write home. Nevertheless, any lump of jelly that has a little convex curvature would mark significant improvements over an open pinhole.

The biggest difference between a good lens and something like the abalone's vitreous mass is this: for best results the lens should be detached from the retina and separated from it by some distance. The gap need not be literally empty. It could be filled by more vitreous mass. What is needed is that the lens should have a higher refractive index than the substance that separates the lens from the retina. There are various ways in which this might be achieved, none of them difficult. I'll deal with just one way, in which the lens is formed by densing up a local region within the front portion of a vitreous mass like that in Figure 5e.

First, remember that a refractive index is something that every transparent substance has. It is a measure of its power to bend rays of light. Human lens-makers normally assume that the refractive index of a lump of glass is uniform through the glass. Once a ray of light has entered a particular glass lens and changed direction appropriately, it goes in a straight line until it hits the other side of the lens. The lens-maker's art lies in grinding and polishing the surface of the glass into precision shapes, and in joining different lenses together in compound cascades.

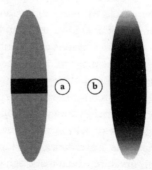

Figure 8 Two kinds of complex lens.

You can glue different kinds of glass together in complicated ways to make compound lenses with lots of different refractive indexes in various parts of them. The lens in Figure 8a, for instance, has a central core made of a different kind of glass with a higher refractive index. But there are still discrete changes from one refractive index to another. In principle, however, there is no reason why a lens should not have a continuously varying refractive index throughout its interior. This is illustrated in Figure 8b. This 'graded index lens' is hard for human lens-makers to achieve because of the way they make their lenses out of glass. But it is easy for living lenses to be built like this because they are not made all at one time: they grow from small beginnings as the young animal develops. And, as a matter of fact, lenses with continuously varying refractive indexes are found in fish, octopuses and many other animals.

Not all lenses evolved by condensing out from a gelatinous mass. Figure 9 shows two insect eyes which form their lenses in quite different ways. These are both so-called simple eyes, not to be confused with the compound eyes which we'll come to in a moment. In the first of these simple eyes, from a cuckoo spit insect or froghopper, the lens forms as a thickening of the cornea – the outer transparent layer. In the second one, from a mayfly, the cornea is not thickened and the lens develops as a mass of colourless, transparent cells. Both these two methods of lens development lend themselves to the same kind of Mount Improbable climb as we've already undertaken for the vitreous mass eye of the worm. Lenses, like eyes themselves, seem to have evolved many times independently. Mount Improbable has many peaks and hillocks.

We should not forget other advanced features of modern eyes such as the apparatuses for changing the focus of an eye,

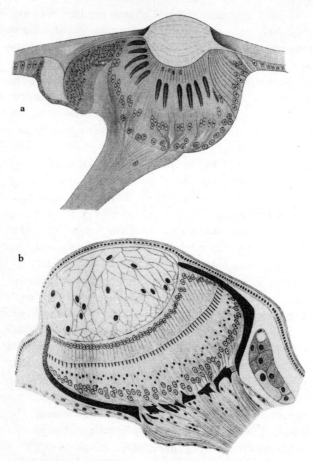

Figure 9 Two different ways for insect lenses to develop:
(a) sawfly larva; (b) mayfly.

for changing the size of the pupil or 'f-stop', and for moving
the eye. There are also all the systems in the brain that are

needed for processing the information from the eye. Moving the eye is important, not just for the obvious reason but, more indispensably, to hold the gaze still while the body moves. Birds do this by using the neck muscles to keep the whole head still, notwithstanding substantial movements of the rest of the body. Advanced systems for doing this involve quite sophisticated brain mechanisms. But it is easy to see that rudimentary, imperfect adjustments would be better than nothing, so there is no difficulty in piecing together an ancestral series following a smooth path up Mount Improbable.

In order to focus rays that are coming from a very distant target, you need a weaker lens than to focus rays that are coming from a close target. To focus sharply both far and near is a luxury one can live without, but in nature every little boost to the chances of survival counts, and as a matter of fact different sorts of animals display a variety of mechanisms for changing the focus of the lens. We mammals do it by means of muscles that pull on the lens and change its shape a little. So do birds and most reptiles. Chameleons, snakes, fishes and frogs do it in the same way a camera does, by pulling the lens a little way forwards or backwards. Animals with smaller eyes don't bother. Their eyes are like a Box Brownie: approximately, though not brilliantly, in focus at all distances. As we get older our eyes sadly become more Box Brownie-like and we often need bifocal glasses to see both near and far.

It is not at all difficult to imagine the gradual evolution of mechanisms for changing focus. When experimenting with the polythene bag filled with water, I quickly noticed that the sharpness of focus could be made better (or worse) by poking the bag with my fingers. Without being consciously aware of the shape of the bag, without even looking at the bag but concentrating on the quality of the image being projected, I simply poked and squashed the bag at random until the focus got better. Any muscle in the vicinity of a lump of vitreous mass could, as a byproduct of contracting for some other purpose, incidentally improve the focus of the lens. This opens up a broad highway for gentle improvement all the way up the slopes of Mount Improbable, which could culminate in either the mammal or the chameleon method of changing the focus.

Changing the aperture – the size of the hole through which light is admitted – may be slightly more difficult, but not much. The reason for wanting to do this is the same as in a camera. For any given sensitivity of film/photocells, it is possible to have too much light (dazzle) as well as too little. Moreover, the narrower the hole, the better the depth of focus – the range of distances that are simultaneously in focus. A sophisticated camera, or eye, has a built in lightmeter which automatically stops down the hole when the sun comes out, and opens up the hole when the sun goes in. The pupil of a human eye is a pretty sophisticated piece of automation technology, something that a Japanese micro-engineer could be proud of.

But, once again, it isn't difficult to see how this advanced mechanism might have got its start on the lower slopes of Mount Improbable. We think of the pupil as circular, but it doesn't have to be. Any shape would do. Sheep and cattle have a long, horizontal lozenge-shaped pupil. So do octopuses and some snakes, but other snakes have a vertical slit. Cats have a pupil which varies from a circle to a narrow, vertical slit:

> Does Minnaloushe know that his pupils
> Will pass from change to change,
> And that from round to crescent,
> From crescent to round they range?

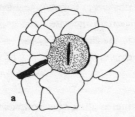

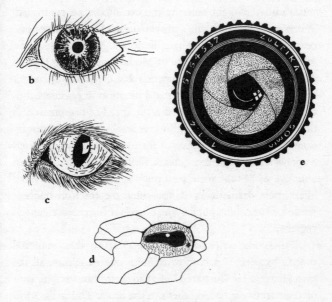

Figure 10 Various pupils including that of a camera. The exact shape of a pupil doesn't matter, which is why it is allowed to be so variable: (a) reticulated python; (b) human; (c) cat; (d) long-nosed tree snake; (e) camera.

Minnaloushe creeps through the grass
Alone, important and wise,
And lifts to the changing moon
His changing eyes.

W. B. Yeats

Even expensive cameras often have a pupil which is a crude polygon rather than a perfect circle. All that matters is that the quantity of light entering the eye should be controlled. When you realize this, the early evolution of the variable pupil ceases to be a problem. There are lots of gentle paths to be followed up the lower slopes of Mount Improbable. The iris diaphragm is no more an impenetrable evolutionary barrier than is the anal sphincter. Perhaps the most important quantity that needs to be improved is the speed of responsiveness of the pupil. Once you have nerves at all, speeding them up is an easy glide up the slopes of the mountain. Human pupils respond fast, as you can quickly verify by shining a torch in your eye while looking at your pupil in a mirror. (You see the effect most dramatically if you shine the torch in one eye while looking at the pupil in the other: for the two are ganged together.)

Retinas, like other parts of the eye, betray their manifold origins by their variable forms. With one exception, all the eyes I have so far illustrated have had their photocells in front of the nerves connecting them to the brain. This is the obvious way to do it, but it is not universal. The flatworm in Figure 1a keeps its photocells apparently the wrong side of their connecting nerves. So does our own vertebrate eye. The photocells point backwards, away from the light. This is not as silly as it sounds. Since they are very tiny and transparent, it doesn't much matter which way they point: most photons will go straight through and then run the gauntlet of pigment-laden baffles waiting to catch them. The only sense in which it even means much to say that the photocells point backwards is that the 'wires' (nerves) connecting them to the brain depart in the wrong direction, towards the light rather than towards the brain. They then run over the front surface of the retina towards one particular place, the so-called 'blind spot'. This is where they dive through the retina into the optic nerve, which is why the retina is blind at this spot. Although we are all technically blind at the spot, we scarcely know it because the brain is so clever at reconstituting the missing bit. We only notice the blind spot if the image of some small discrete object, which we have independent evidence exists, moves on to it: it then appears to go out like a light, apparently replaced by the general background colour of the area.

I've said that it makes little difference if the retina is back to front. A case could be made that, absolutely all other things being equal, it might have been better if our retinas were the right way round. It is perhaps a good example of the fact that Mount Improbable has more than one peak, with deep valleys between. Once a good eye has started to evolve with its retina back to front, the only way to ascend is to improve the present design of eye. Changing to a radically different design involves going downhill, not just a little way but down a deep chasm, and that is not allowed by natural selection. The vertebrate retina faces the way it does because of the way it develops in the embryo, and this certainly goes back to its ancient ancestors. The eyes of many invertebrates develop in different ways, and their retinas are consequently the 'right' way round.

Setting aside the interesting fact of their pointing backwards, vertebrate retinas scale some of the loftiest peaks on the mountain. The human retina has about 166 million photocells, divided into various kinds. The basic division is into rods (specialized for low-precision, non-colour vision at relatively low light levels) and cones (specialized for high-precision colour vision in bright light). As you read these words you are using only cones. If Juliet saw Halley's Comet, it would have been her rods that were responsible. The cones are concentrated in a small central area, the fovea (you are reading with your foveas), where there are no rods. This is why, if you want to see a really dim object like Halley's Comet, you must point your eyes not directly at it but slightly away, so that its meagre light is off the fovea. Numbers of photocells, and differentiation of photocells into more than one type, present no special problems from the point of view of climbing Mount Improbable. Both kinds of improvement obviously constitute smooth gradients up the mountain.

Big retinas are better than small retinas. You can fit more photocells in, and you can see more detail. As always, there are costs. But there is a way in which a small animal can, in effect, enjoy a larger retina than it pays for. Professor Michael Land of Sussex University, who has an enviable track record for exotic discoveries in the world of eyes and from whom I have learned much of what I know about animal eyes, found a fascinating example in jumping spiders. These engaging little animals, whose habit of cocking their heads to look at you gives them an almost human charm, stalk their prey like a cat and then jump on to it explosively and without warning. Explosive it more or less literally is, by the way, for they jump by hydraulically pumping fluid into all eight legs simultaneously – a little like the way we (those of us who have them) erect our penises, but their 'leg erections' are sudden rather than gradual. No spiders have compound eyes: jumping spiders have taken the camera eye up to a remarkable peak of economy. What Land discovered was an extraordinary retina. Instead of being a wide sheet on which a full image can be projected, it is a long, vertical strip, not wide enough to accommodate a whole image. But the spider makes up for the narrowness of its retina in a dynamic way. It moves its retina systematically about, 'scanning' the area where an image might be projected. Its effective retina is, therefore, much larger than its actual retina. If the jumping spider's retina finds an interesting object, like a moving fly or another jumping spider, it concentrates its scanning movements in the precise area of the target. This gives it the dynamic equivalent of a fovea. Using this clever trick, jumping spiders have carried the lens eye to a respectable little peak in their local area of Mount Improbable.

I introduced the lens as an excellent remedy for the shortcomings of the pinhole eye. It isn't the only one. A curved mirror constitutes a different principle from a lens, but it is a good alternative solution to the same problem of gathering a large amount of light from each point on an object, and focusing it to a single point on an image. For some purposes a curved mirror is actually a more economical solution to the problem than a lens, and the biggest optical telescopes in the world are all reflectors (Figure 11a). A minor problem with a reflecting telescope is that the image is formed in front of the mirror, actually in the pathway of the incoming rays. Reflecting telescopes usually have a small mirror to reflect the focused image sideways into an eyepiece or a camera. The small mirror doesn't get in the way, not enough to spoil the image, anyway. No focused image of the little mirror is seen: it merely causes a small reduction in the total amount of light hitting the big mirror at the back of the telescope.

The curved mirror, then, is a theoretically workable physical solution to an important problem. Are there any examples of curved mirror eyes in the animal kingdom? The earliest suggestion along these lines was made by my old Oxford Professor, Sir Alister Hardy, commenting on his painting of a remarkable deep-sea crustacean called *Gigantocypris* (Figure 11b). Astronomers capture what few photons arrive from distant stars with huge curved mirrors in observatories like Mount Wilson and Palomar. It is tempting to think that *Gigantocypris* is doing the same thing with the few photons that penetrate the deep oceans, but recent investigations by Michael Land rule out any resemblance in detail. It is at the moment not clear how *Gigantocypris* sees.

There is another kind of animal, however, that definitely uses a *bona fide* curved mirror to form an image, albeit it has a lens as well to help. Once again, it was discovered by that King Midas of animal eye research, Michael Land. The animal is the scallop. The photograph in Figure 11c is an enlargement of a small piece (two shell-corrugations in width) of the gape of one of these bivalves. Between the shell and the tentacles is a row of dozens of little eyes. Each eye forms an image, using a curved mirror which lies well behind the retina. It is this mirror that causes each eye to glow like a tiny blue or green pearl. In section, the eye looks like Figure 11d. As I mentioned, there is a lens as well as a mirror, and I'll come back to this. The retina is the whole greyish area lying between the lens and the curved mirror. The part of the retina

---

Figure 11 (opposite) Curved mirror solutions to the problem of forming images: (a) reflecting telescope; (b) *Gigantocypris*, a large planktonic crustacean painted by Sir Alister Hardy; (c) scallop: eyes peeping through gap in shell; (d) cross-section of scallop eye; (e) Cartesian oval.

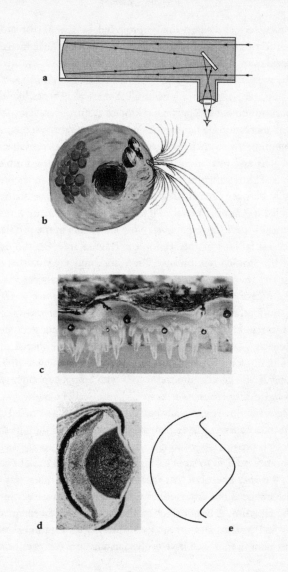

which sees the sharp image projected by the mirror is the portion tightly abutting the back of the lens. That image is upside down and it is formed by rays reflected backwards by the mirror.

So, why is there a lens at all? Spherical mirrors like this one are subject to a particular kind of distortion called spherical aberration. A famous design of reflecting telescope, the Schmidt, overcomes the problem by a cunning combination of lens and mirror. Scallop eyes seem to solve the problem in a slightly different way. Spherical aberration can theoretically be overcome by a special kind of lens whose shape is called a 'Cartesian oval'. Figure 11e is a diagram of a theoretically ideal Cartesian oval. Now look again at the profile of the actual lens of the scallop eye (Figure 11d). On the basis of the striking resemblance, Professor Land suggests that the lens is there as a corrector for the spherical aberration of the mirror which is the main image-forming device.

As for the origin of the curved mirror eye on the lower slopes of its region of the mountain, we can make an educated guess. Reflecting layers behind retinas are common in the animal kingdom, but for a different purpose, not image forming as in scallops. If you go out into the woods with a bright spotlight you will see numerous twin beams glaring straight back at you. Many mammals, especially nocturnal ones, have a so-called tapetum, a reflecting layer behind the retina. What the tapetum does is provide a second chance of catching photons that the photocells failed to stop: each photon is reflected straight back to the very photocell that missed it coming the other way, so the image is not distorted. Invertebrates, too, have discovered the tapetum. A bright torch in the woods is an excellent way to find certain kinds of spider. Tapetums for capturing every last photon may well have evolved in ancestral cup eyes before lenses. Perhaps the tapetum is the 'preadaptation' which, in a few isolated creatures, has become modified to form a reflecting telescope kind of eye. Or the mirror may have arisen from another source. It is hard to be sure.

Lenses and curved mirrors are two ways of sharply focusing an image. In both cases the image is upside down and left–right reversed. A completely different kind of eye, which produces an image the right way up, is the compound eye, favoured by insects, crustaceans, some worms and molluscs, king crabs (strange marine creatures said to be closer to spiders than to real crabs) and the large group of now extinct trilobites. Actually there are several different kinds of compound eye. I'll begin with the most elementary kind, the so-called 'apposition compound' eye. To understand how the apposition eye works, we go back nearly to the bottom of Mount Improbable. As we have seen, if you want an eye to see an image or indeed go beyond signalling the mere intensity of light, you need more than one photocell, and they must pick up light from different directions. One way to make them look in different directions is to place them in a cup, backed by an opaque screen. All the eyes we have so far talked about have been descendants of this concave cup principle. But perhaps an even more obvious solution to the problem is to place the photocells on the convex, outside surface of a cup, thereby causing them to look outwards in different directions. This is a good way to think of a compound eye, at its simplest.

Remember when we first introduced the problem of forming an image of a dolphin. I pointed out that the problem could be regarded as the problem of having too many images. An infinite number of 'dolphins' on the retina, every way up and in every position on the retina adds up to no visible dolphin at all. The pinhole eye worked because it filtered out almost all the rays, leaving only the minority that cross each other in the pinhole and form a single upside-down image of the dolphin. We treated the lens as a more sophisticated version of the same principle. The apposition compound eye solves the problem in an even simpler way.

The eye is built as a dense cluster of long straight tubes, radiating out in all directions from the roof of a dome. Each tube is like a gunsight which sees only the small part of the world in its own direct line of fire. In terms of our filtering metaphor, we could say that rays coming from other parts of the world are prevented, by the walls of the tube and the backing of the dome, from hitting the back of the tube where the photocells are.

That's basically how the apposition compound eye works. In practice, each of the little tube eyes, called an ommatidium

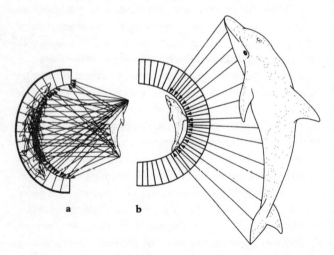

a   b

Figure 12 (a) reproduction of Figure 3; (b) the cup turned inside out. Principle of the apposition compound eye.

(plural ommatidia), is a bit more than a tube. It has its own private lens, and its own tiny 'retina' of, usually, half a dozen or so photocells. Insofar as each ommatidium produces an image at all at the bottom of the narrow tube, that image is upside-down: the ommatidium works like a long, poor-quality, camera eye. But the upside-down images of the individual ommatidia are ignored. The ommatidium reports only how much light comes down its tube. The lens serves only to gather more light rays from the ommatidium's gunsight direction and focus them onto the retina. When all the ommatidia are taken together, their summed 'image' is the right way up, as shown in Figure 12b.

As always, 'image' doesn't have to mean what we humans would think of as an image: an accurate, Technicolor perception of an entire scene. Instead, we are talking about *any* kind of ability to use the eyes to distinguish what is going on in different directions. Some insects might, for example, use their compound eyes only to track moving targets. They might be blind to still scenes. The question of whether animals see things in the same way as we do is partly a philosophical one and it may be a more than usually difficult task trying to answer it.

The compound eye principle works well enough for, say, a dragonfly zeroing in on a moving fly but, in order for a compound eye to see as much detail as we see, it would need to be hugely bigger than the kind of simple camera eye that we possess. Here is approximately why this is. Obviously, the more ommatidia you have, all looking in slightly different directions, the more fine detail you can see. A dragonfly may have 30,000 ommatidia and it is pretty good at hawking insects on the wing. But in order to see as much detail as we can see, you'd need millions of ommatidia. The only way to fit in millions of ommatidia is to make them exceedingly tiny. And unfortunately there is a strict limit on how small an ommatidium can be. It is the same limit as we met in talking about very small pinholes, and it is called the diffraction limit. The consequence is that, in order to make a compound eye see as precisely as the human camera eye, the compound eye would have to be ludicrously large: twenty-four metres in diameter! The German scientist Kuno Kirschfeld dramatized this by drawing what a man would look like if he could see as well as a normal man can see, but using compound eyes (Figure 13). The honeycomb pattern on the drawing is

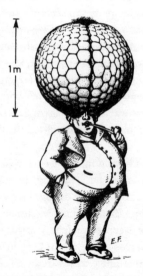

Figure 13 Kuno Kirschfeld's picture of how a man with compound eyes would look if he wanted to see as well as a normal human.

impressionistic, by the way. Each facet drawn actually stands for 10,000 ommatidia.

The reason the man's compound eyes are only about one metre across instead of twenty-four is that Kirschfeld made allowance for the fact that we humans only see very precisely in the centre of our retina. He took an average of our precise central vision and our much less precise vision towards the edges of our retina, and came up with the one metre eye shown. Whether one metre across or twenty-four, a compound eye this large is impractical. The moral is, if you want to see precise, detailed images of the world, use a simple camera eye with a single, good lens, not a compound eye. The Swedish biologist Dan Nilsson even remarks of compound eyes that 'It is only a small exaggeration to say that evolution seems to be fighting a desperate battle to improve a basically disastrous design.'

Why, then, don't insects and crustaceans abandon the compound eye and evolve camera eyes instead? It may be one of those cases of becoming trapped the wrong side of a valley on the *massif* of Mount Improbable. To change a compound eye into a camera eye, there has to be a continuous series of workable intermediates: you cannot travel down into a valley as a prelude to mounting a higher peak. So, what is the problem about intermediates between a compound eye and a camera eye?

At least one outstanding difficulty comes to mind. A camera eye forms an upside-down image. A compound eye's image is the right way up. Finding an intermediate between those two is a tough proposition, to put it mildly. A possible intermediate is no image at all. There are some animals, living in the deep sea or otherwise in near total darkness, who have so few photons to play with that they give up on images altogether. All that they can hope for is to know whether there is light at all. An animal such as this could lose its image-processing nervous apparatus altogether and hence be in a position to make a fresh start up a completely different slope of the mountain. It could therefore constitute an intermediate on the path from a compound eye to a camera eye.

Some deep-sea crustaceans have large compound eyes but no lenses or optical apparatus at all. Their ommatidia have lost their tubes and their photocells are exposed right at the outer surface of where they will pick up what few photons there are, regardless of direction. From there it would seem but a small step to the remarkable eye of Figure 14. It belongs to a crustacean, called *Ampelisca*, which doesn't live particularly deep – perhaps it is on the way back up again from deep sea ancestors. *Ampelisca*'s eye works as a camera eye, with a single lens forming an upside-down image on a shallow cup retina. But the retina is clearly derived from a compound eye and consists of the remains of a bank of ommatidia. A small

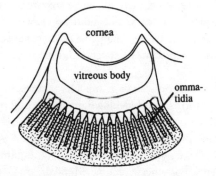

Figure 14 A camera eye with a compound eye in its ancestral history. The remarkable eye of *Ampelisca*.

step, maybe, but only if, during the interregnum of near total blindness, the brain had enough evolutionary time to 'forget' all about processing right-way-up images.

That is an example of evolution from compound eye to camera eye (yet another example, by the way, of the ease with which eyes seem to evolve independently all around the animal kingdom). But how did the compound eye evolve in the first place? What do we find on the lower slopes of this particular peak of Mount Improbable?

Once again we may be helped by looking around the modern animal kingdom. Outside the arthropods (insects, crustaceans and their kin), compound eyes are found only in some Polychaete worms (bristleworms and tubeworms) and in some bivalve molluscs (again, presumably independently evolved). The worms and molluscs are helpful to us as evolutionary historians because they also include among their number some primitive eyes which look like plausible intermediates strung out along the lower slopes of Mount Improbable leading to a compound eye peak. The eyes in Figure 15 come from two different worm species. Once again, these are not ancestors, they are modern species and they are probably not even descended from the true intermediates. But they could easily be giving us a glimpse of what the evolutionary progression might have been like, from a loose clustering of photocells on the left to a proper compound eye on the right. This slope is surely just as gentle as the one we strolled up to reach the ordinary camera eye.

Ommatidia, as we have so far discussed them, depend for their effectiveness on being isolated from their neighbours. The gunsight that is looking at the dolphin's tailtip must not pick up rays from other parts of the dolphin, or we shall be back with our original problem of millions of dolphin images. Most ommatidia achieve isolation by having a sheath of dark

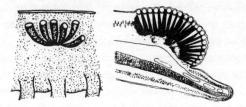

Figure 15 Possibly primitive compound eyes from two kinds of worms.

pigment around the tube. But there are times when this has bad side effects. Some sea creatures rely on transparency for their camouflage. They live in sea water and they look like sea water. The essence of their camouflage, then, is to avoid stopping photons. Yet the whole point of dark screens around ommatidia is to stop photons. How to escape from this cruel contradiction?

There are some deep-sea crustaceans who have come up with an ingenious partial solution. They don't have screening pigment, and their ommatidia are not tubes in the ordinary sense. Instead, they are transparent light guides, working just like man-made fibre-optic systems. Each light guide swells, at its front end, into a tiny lens, of varying refractive index like a fish eye. Lens and all, the light guide as a whole concentrates a large amount of light on to the photocells at its base. But this includes only light coming from straight in the line of the gunsight. Beams coming sideways at a tube, instead of being shielded by a pigment screen, are reflected back and don't enter the light guide.

Not all compound eyes even try to isolate their private supply of light. It is only eyes of the apposition type that do. There are at least three different kinds of 'superposition' compound eye which do something more subtle. Far from trapping rays in tubes or fibre-optic light guides, they allow rays that pass through the lens of one ommatidium to be picked up by a neighbouring ommatidium's photocells. Far from there being a tube connecting the front to the back end of an ommatidium, some types of superposition compound eye have an empty, transparent zone, shared by all omma-tidia. The lenses of all ommatidia conspire together to form a single image on a shared retina which is jointly put together from the light sensitive cells of all ommatidia.

The image in a superposition compound eye, like that of apposition compound eyes but unlike that of camera eyes or that of Figure 14's *Ampelisca*, is the right way up. This is what you'd expect, assuming that superposition eyes are derived from apposition ancestors. It makes historical sense, and it must have made for an effortless transition as far as the brain was concerned. But it is still a very remarkable fact. For consider the physical problems of constructing a single right-way-up image in this way. Each individual ommatidium in an appo-sition eye has a normal lens in front of it which, if it makes an image at all, makes an upside down one. To convert an apposition eye into a superposition one, therefore, the rays, as they pass through each lens, have somehow to be turned the right way up. Not only this, all the separate images from the different lenses have to be carefully superimposed to give one shared image. The advantage of doing this is that the shared image is much brighter. But the physical difficulties of turning the rays round are formidable. Amazingly, not only has the problem been solved in evolution, it has been solved in at least three independent ways: using fancy lenses, using fancy mirrors, and using fancy neural circuitry. The details are so intricate that to spell them out would unbalance this already quite complicated chapter, and I'll deal with them only briefly.

A single lens turns the image upside down. By the same token another lens, a suitable distance behind it, would turn it the right way up again. The combination is used in an instrument called a Keplerian telescope. The equivalent effect can be achieved in a single complex lens, using fancy gradations of refractive index. As we have seen, living lenses, unlike man-made ones, are good at achieving gradations of refractive index. I won't attempt to go into the details, but this method of simulating the effect of a Keplerian telescope is used by mayflies, lacewings, beetles, moths, caddises, and members of five different groups of crustaceans. The distance of their cousinship suggests that at least several of these groups evolved the same Keplerian trick independently of one another. An equivalent trick is pulled by three groups of crustaceans, using mirrors. Two of these three groups also contain members that do the lens trick. Indeed, if you look at which animal groups have adopted which of the several different kinds of compound eye, you notice a fascinating thing. The different solutions to problems pop up here, there and everywhere, suggesting, yet again, that they evolve rapidly and at the drop of a hat.

'Neural superposition' or 'wired-up superposition' has evolved in the large and important group of two-winged insects, the flies. A similar system occurs in water boatmen, where it seems to have evolved – yet again – independently. Neural superposition is fiendishly cunning. In a way it shouldn't be called superposition at all, because the ommatidia are isolated tubes just as in apposition eyes. But they achieve a superposition-like effect by ingenious wiring of nerve cells behind the ommatidia. Here's how. You'll remember that the 'retina' of a single ommatidium is made up of about half a dozen photocells. In ordinary apposition eyes, the firing of all six photocells is simply added together, which is why I put 'retina' in quotation marks: all photons that shoot down the tube are counted, regardless of which photocell they hit. The only point of having several photocells is to increase the total sensitivity to light. This is why it doesn't matter that the tiny image at the bottom of an apposition ommatidium is technically upside-down.

But in the eye of a fly the outputs of the six cells are not pooled *with each other*. Instead, each one is pooled with the outputs of *particular* cells from *neighbouring* ommatidia (Figure 16). In the interests of clarity, the scale is all wrong in this diagram. For the same reason, the arrows don't represent rays (which would be bent by the lenses) but mappings from points on the dolphin to points in the bottoms of tubes. Now, see the shattering ingenuity of this scheme. The essential idea is that those photocells that are looking at the head of the dolphin in one ommatidium are ganged together with those photocells that are looking at the head of the dolphin in neighbouring ommatidia. Those photocells that are looking at the tail of the dolphin in one ommatidium are wired up together with those photocells that are looking at the tail of the dolphin in neighbouring ommatidia. And so on. The result is that each bit of the dolphin is being signalled by a larger number of photons than there would be in an ordinary apposition eye with a simple tube arrangement. It is a kind of computational, rather than an optical, solution to our old problem of how to augment the number of photons arriving from any one point on our dolphin.

You can see why this is called superposition, even though it strictly isn't superposition. In true superposition, using

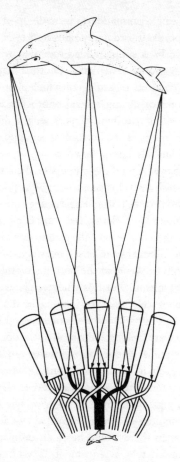

Figure 16 The ingenious principle of the 'wired-up super-position' compound eye.

fancy lenses or mirrors, light coming through neighbouring facets is superimposed so that photons from the dolphin's head end up in the same place as other photons from the dolphin's head; photons from the tail end up in the same place as other photons from the tail. In neural superposition, the photons still end up in different places, as they would in an apposition eye. But the signal *from* those photons ends up in the same place, due to the artful plaiting of the wires leading to the brain.

The same Dan Nilsson, working with his colleague Susanne Pelger, used a computer model to estimate how long it might take to evolve a camera eye with a lens from a flat retina without a lens. The answer was the astonishingly short time of less than half a million generations. By geological standards this is more or less instantaneous. You'd be lucky to find fossils that recorded the transitional stages. Exact estimates have not been done for compound eyes or any of the other designs of eye, but I doubt if they'd be significantly slower. One doesn't ordinarily expect to be able to see the details of eyes in fossils, because their bits are too soft to fossilize. Compound eyes are an exception because much of their detail is betrayed in the elegant array of more or less horny facets on the outer surface. A trilobite eye, from the Devonian era nearly 400 million years ago, looks just as advanced as a modern compound eye. This is what we should expect if the time it takes to evolve an eye is negligible by geological standards.

A central message here is that eyes evolve easily and fast, at the drop of a hat. I began by quoting the conclusion of one authority that eyes have evolved independently at least forty times in different parts of the animal kingdom. On the face of it, this message might seem challenged by an intriguing set of experimental results, recently reported by a group of workers in Switzerland associated with Professor Walter Gehring. I shall briefly explain what they found, and why it does not really challenge our conclusion. Before I begin, I need to apologize for a maddeningly silly convention adopted by geneticists over the naming of genes. The gene called *eyeless* in the fruitfly *Drosophila* actually makes eyes! Wonderful, isn't it? The reason for this wantonly confusing piece of terminological contrariness is actually quite simple, and even rather interesting. We recognize what a gene does by noticing what happens when it goes wrong. There is a gene which, when it goes wrong (mutates), causes flies to have no eyes. The position on the chromosome of this gene is therefore named the *eyeless* locus ('locus' is the Latin for 'place' and it is used by geneticists to mean a slot on a chromosome where alternative forms of a gene sit). But usually when we speak of the locus named *eyeless* we are actually talking about the normal, undamaged form of the gene at that locus. Hence the paradox that the *eyeless* gene makes eyes. It is like calling a loudspeaker a 'silence device' because you have discovered that, when you take the loudspeaker out of a radio, the radio is silent. I shall have none of it. I am tempted to rename the gene *eyemaker*, but this would be confusing too. I shall certainly not call it *eyeless* and shall adopt the recognized abbreviation *Ey*.

Now, it is a general fact that although all of an animal's genes are present in all its cells, only a minority of those genes are actually turned on or 'expressed' in any given part of the body. This is why livers are different from kidneys, even though both contain the same complete set of genes. In the adult *Drosophila, Ey* usually expresses itself only in the head, which is why the eyes develop there. George Halder, Patrick Callaerts and Walter Gehring discovered an experimental manipulation that led to *Ey*'s being expressed in other parts of the body. By doctoring *Drosophila* larvae in cunning ways, they succeeded in making *Ey* express itself in the antennae, the wings and the legs. Amazingly, the treated adult flies grew up with fully formed compound eyes on their wings, legs, antennae and elsewhere. Though a bit smaller than ordinary eyes, these 'ectopic' eyes are proper compound eyes with plenty of properly formed ommatidia. They even work. Well, we don't know that the flies actually see anything through them, but electrical recording from the nerves at the base of the ommatidia shows that they are sensitive to light.

That is remarkable fact number one. Fact number two is even more remarkable. There is a gene in mice called *Small Eye* and one in humans called *Aniridia*. These, too, are named using the geneticists' negative convention: mutational damage to these genes causes reduction or absence of eyes or bits of eyes. Rebecca Quiring and Uwe Waldorf, working in the same Swiss laboratory, found that these particular mammal genes are almost identical, in their DNA sequences, to the *Ey* gene in *Drosophila*. In other words, the same gene has come down from remote ancestors to modern animals as distant from each other as mammals and insects. Moreover, in both these major branches of the animal kingdom it seems to have a lot to do with eyes. Remarkable fact number three is almost too startling. Halder, Callaerts and Gehring succeeded in introducing the mouse gene into Drosophila embryos. *Mirabile dictu*, the mouse gene succeeded in inducing ectopic eyes on *Drosophila* legs. It is an insect compound eye that is induced, by the way, not a mouse eye. The mouse gene has simply switched on the eyemaking developmental machinery of *Drosophila*. Genes with pretty much the same DNA sequence as *Ey* have been found also in molluscs, marine worms called nemertines, and seasquirts (bag-shaped marine invertebrates remotely related to vertebrates). *Ey* may very well be universal among animals, and it may turn out to be a general rule that a version of the gene taken from a donor in one part of the animal kingdom can induce eyes to develop in recipients in an exceedingly remote part of the animal kingdom.

What does this spectacular series of experiments mean for our conclusion? Were we wrong to think that eyes have developed forty times independently? I don't think so. At least the spirit of the statement that eyes evolve easily and at the drop of a hat remains unscathed. These experiments probably do mean that the common ancestor of *Drosophila*, mice, humans, seasquirts and so on had eyes. The remote common ancestor had vision of some kind, and its eyes, whatever form they may have taken, probably developed under the influence of a sequence of DNA similar to modern *Ey*. But the actual form of the different kinds of eye, the details of retinas and lenses or mirrors, the choice of compound versus simple, and if compound the choice among apposition or various kinds of superposition, all these evolve independently and rapidly. We know this by looking at the sporadic – almost capricious – distribution of all these various devices and systems, dotted around the animal kingdom. In brief, animals often have an eye that resembles their remoter cousins more than it resembles their closer cousins. The conclusion remains unshaken by the demonstration that the common ancestor of all these animals probably had eyes of some kind, and that the embryonic development of all eyes seems to have enough in common to be inducible by the same DNA sequence.

After Michael Land had kindly read and criticized the first draft of this essay, I invited him to attempt a visual representation of the eye region of Mount Improbable, and Figure

17 shows what he drew. It is in the nature of metaphors that they are good for some purposes but not others, and we must be prepared to modify them, or even drop them altogether, when necessary. This is not the first occasion when the reader will have noticed that Mount Improbable, for all that it has a singular name like the Jungfrau, is actually a more complicated, multiple peaked affair.

That other great authority on animal eyes, Dan Nilsson, who also read this text in draft, summed up the central message by calling my attention to what may be the most bizarre example of the *ad hoc* and opportunistic evolution of an eye. Three times independently, in three different groups

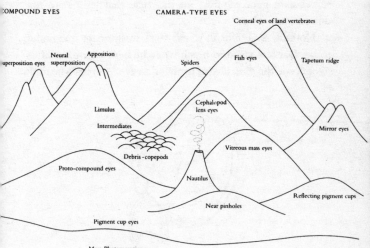

Figure 17 The eye region of the Mount Improbable range: Michael Land's landscape of eye evolution.

of fish, the so-called 'four-eyed' condition has evolved. Probably the most remarkable of the four-eyed fish is *Bathylychnops exilis*. It has a typical fish eye looking outwards in the usual direction. But a secondary eye has evolved in addition, lodged in the wall of the main eye and looking straight downwards. What it looks at, who knows? Perhaps *Bathylychnops* suffers from a terrible predator with the habit of approaching from below. From our point of view the interesting thing is this. The embryological development of the secondary eye is completely different from that of the main eye, although we may surmise that its development may turn out to be induced in nature by a version of the *Ey* gene. In particular, as Dr Nilsson put it in his letter to me, 'This species has re-invented the lens despite the fact it already had one. It serves as a good support for the view that lenses are not difficult to evolve.'

Nothing is as difficult to evolve as we humans imagine it to be. Darwin gave too much when he bent over backwards to concede the difficulty of evolving an eye. And his wife took

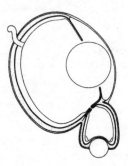

Figure 18 A remarkable double eye, that of the fish *Bathylychnops exilis*.

too much when she underlined her scepticism in the margin. Darwin knew what he was doing. Creationists love the quotation that I gave earlier, but they never complete it. After making his rhetorical concession, Darwin went on:

When it was first said that the sun stood still and the world turned round, the common sense of mankind declared the doctrine false; but the old saying of *Vox populi, vox Dei*, as every philosopher knows, cannot be trusted in science. Reason tells me, that if numerous gradations from an imperfect and simple eye to one perfect and complex, each grade being useful to its possessor, can be shown to exist, as is certainly the case; if further, the eye ever slightly varies, and the variations be inherited, as is likewise certainly the case; and if such variations should ever be useful to any animal under changing conditions of life, then the difficulty of believing that a perfect and complex eye could be formed by natural selection, though insuperable by our imagination, cannot be considered real.

Human brains, though they sit atop one of its grandest peaks, were never designed to imagine anything as slow as the long march up Mount Improbable.

POCKET PENGUINS